Bibliographic information published by the German National Library:

The German National Library lists this publication in the National Bibliography; detailed bibliographic data are available on the Internet at http://dnb.dnb.de .

Imprint:

Print and binding: Books on Demand GmbH, Norderstedt Germany
ISBN: 9783668516724

This book at GRIN:

http://www.grin.com/en/e-book/373854/a-comparative-analysis-of-the-employment-situation-in-bangladesh-potential

Mahir Faysal

A Comparative Analysis of the Employment Situation in Bangladesh. Potential Sectors and Required Policies for Employment

GRIN Publishing

Abstract

Bangladesh is a country with 163 million people where employment is the major source of income and livelihood for most of the economic active people. The labor force is growing in Bangladesh at higher rate than the rate of growth in population. In 2015 Government of Bangladesh claimed that the labor force participation rate was 57.1 percent with 4.3 percent unemployment rate which was about 2.6 million in number. As the country is experiencing its first demographic dividend these young population demand both quality and quantity opportunities in the employment sector. The main objective of the paper was to provide an understanding of employment situation of Bangladesh. The study found the vulnerable situation of job opportunity and underemployment with gender and region specific discrimination. The findings also showed that small and medium enterprises (Smes), ICT Sector, vocational training and overseas employment and employment in RMG sector were potentials sectors of employment opportunities for Bangladesh.

Table of contents

1. Introduction

Employment is a major source of income and livelihood for most of the people of Bangladesh. According to PRB (2016) Bangladesh is a country of 163 million people in which 12.9 % of the total population live in extreme poverty (World Bank 2016).This poverty & employment are highly associated with each other. Poverty reduces with betterment in employment. On the other hand employment creates opportunity & capability to purchase better goods and services. That's why employment issue has been included in MDGs, SDGs & also in government policies of Bangladesh for the development of the country. One of the main development challenges for Bangladesh is providing enough jobs both in number and quality for the new entrants into the labor force and who are experiencing underemployments. The labor force is growing in Bangladesh at higher rate than the rate of growth in population. This is likely to persist over the next two decades or longer because the country is going through of its first demographic dividend. Bangladesh is experiencing third stage of demographic dividend with its huge youth bulge as a result of declining population growth & transition from high stationary to late expanding. According to Bangladesh Bureau of Statistics (2015), the labor force participation rate is only 57.1% with 4.3% unemployment rate which is about 2.6 million in number where female are more vulnerable 7.3% than male 3%. Still 54.9 % of the economically active population is related with non-agricultural job sector. Data also shows Bangladesh is experiencing 20.3% underemployment rate. Providing quality jobs for this huge population is a challenging issue. This paper seeks to provide an understanding of employment situation of Bangladesh analyzing potential sectors and required policies for employment.

1.1. Methodology and Data Sources

The study follows an analytical research. Data were collected form secondary sources. Sources of data were several government conducted survey specially "Labour Force Survey 2013"with different published journals, books and reports. Data were also collected from different concerned non-government organizations. This paper has used simple statistical tables and graphs to analyze the issues.

2. Demographic Dividend and Economic Active Population in Bangladesh

According to BBS (2011) age distribution of Bangladesh shows that the largest reproductive segments are 10-14 &15-19 which refers to the adolescents group according to UN definition (10-19 aged group) contributes about 19% of the total population of the country. The population pyramid of the country shows that Bangladesh is experiencing youth bulge as a result of declining population growth & transition from high stationary to late expanding (3rd stage of demographic transition) where it has already entered into demographic dividend because the data was taken from 2011 census. Whatever, it will last till 2033 (Alam, 2012). According to UNFPA (2014) 47.6 million or 30 percent of the total 158.5 million people in Bangladesh were young (10-24 years. This means that Bangladesh needs to invest right now transforming these youth economic active population into human resource if it wants to reap the benefits of demographic dividend. Nabi (2011) said that this demographic opportunity can bring a huge dividend, if this labour force is converted into resource. As this window of opportunity opens for a population only once of its history. If Bangladesh fails to grab this demographic opportunity as early as possible, this huge number of economic active population can create a disastrous hazard for the nation.

2.1. Labor Force Characteristics and Employment Situation of Bangladesh

Small size of the employment generation with a huge number of population is the main problem for Bangladesh. It has experienced a long structural economic change since its independence. The country has moved from an agrarian to an industry and services sector based economy gradually. Although the country is experiencing demographic dividend but the number of unemployed people has increased over the years.

Table 1 shows that the economically active population of the country is increasing day by day. In 2002-03 it was 46.3 million and in 2013 it is 60.7 million. Data shows that women were less in number among economically active population in Bangladesh. Table also shows that 58.1 million population are employed where women were again lag behind. Refined activity rate of labor force participation rate of the country shows that 57.1 % of economically active population is participating in labor force. Among male it is 81.7 percent and among female 32.9 percent.

Table 1 Characteristics of labor force in Bangladesh

Category	2002-03	2005-06	2010	2013
Economically active population in million (15+)				
Bangladesh	46.3	49.5	56.7	60.7
Male	36	37.3	39.5	42.5
Female	10.3	12.1	17.2	18.2
Employed population in million (15+)				
Bangladesh	44.3	47.4	54.1	58.1
Male	34.5	36.1	37.9	41.2
Female	9.8	11.3	16.2	16.8
Labor force participation rate (%) (refined activity rate)				
Bangladesh	57.3	58.5	59.3	57.1
Male	87.4	86.8	82.5	81.7
Female	27.4	27.4	34.5	33.5

Source: Bangladesh Bureau of Statistics, Labor Force Surveys 2013

As shown in figure 1 the proportion of population aged 15 and above who were currently economically active in 2013 was 57.1 percent. By sex the rate was higher among male as mentioned earlier and by region the rate is slightly higher in rural area than urban.

Figure 1 Labor force participation rate by category

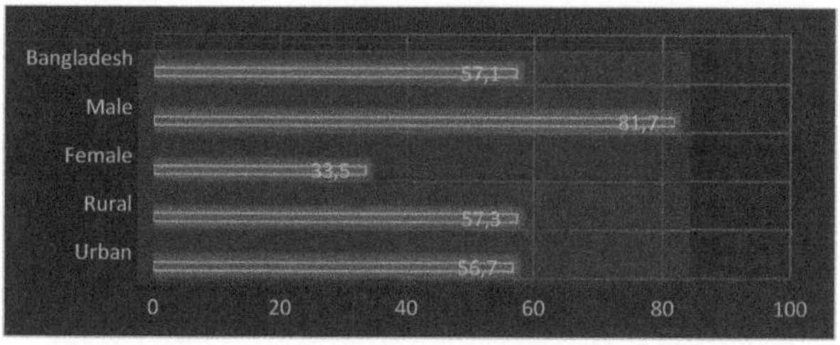

Source: Bangladesh Bureau of Statistics, Labor Force Surveys 2013

Table 2 shows the employed population by broad economic sector. The largest population of the country was involved in agriculture 45.1 percent followed by 34.1 percent in service and 20.8 in industrial sector. More than half of the rural population was involved in agriculture 56.2 percent and more than half of the urban population was involved in service sector.

Table 2 Employed population by broad economic sector (%)

Economic sector	Rural	Urban	Bangladesh
Agriculture	56.2	16.2	45.1
Industry	17.8	28.5	20.8
Service	25.9	55.3	34.1

Source: Bangladesh Bureau of Statistics, Labor Force Surveys 2013

Figure 2 illustrates the percentage proportion of employed person by years in broad economic sector. The proportion of population who were involved in agricultural sectors is decreasing overtime where people involving in nonagricultural sector is increasing in Bangladesh.

Figure 2 Proportion of employed person by economic sector (%)

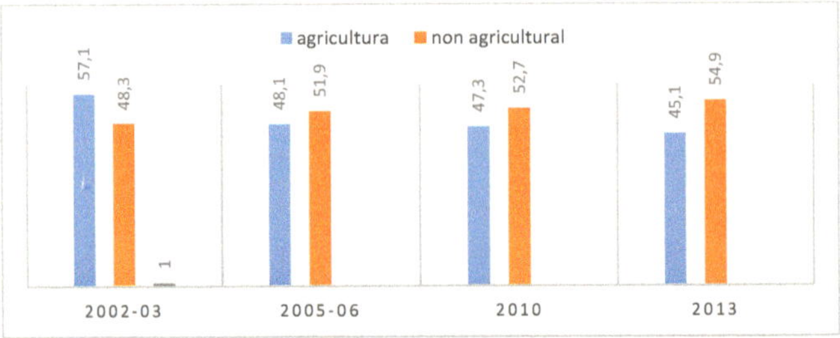

Source: Bangladesh Bureau of Statistics, Labor Force Surveys 2013

In Bangladesh 87.4 percent were involved in informal sector. Female were large in number than male. Contribution of informal sector was higher in rural area 92.8 percent than urban 75.1 percent. Urban population has most involvement in formal employment which was about two times higher than national figure.

Figure 3 Employed population by formal & informal employment (%)

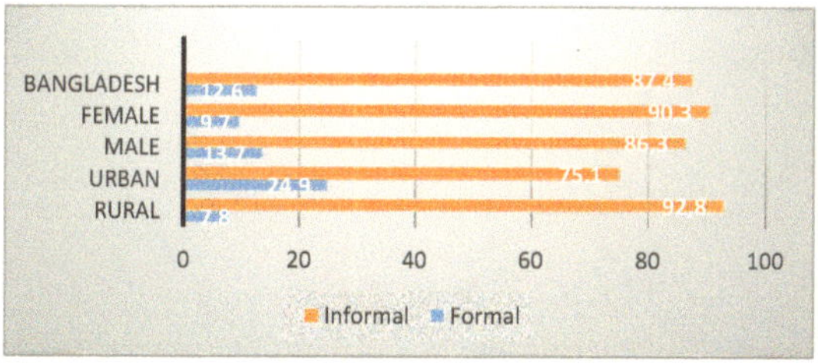

Source: Bangladesh Bureau of Statistics, Labor Force Surveys 2013

Table 3 indicates that largest share of the employed population is involved in skilled agricultural, forestry & fisheries sector where women are more involved than male. 18.1 percent are involved in Craft & related trade sector then Service and sell sector 19.7 percent where male are more in number than female in this sector . 13.5 percent are also involved in Elementary occupations.

Table 3: Employed population aged 15 and above by occupations

Occupations	Male	Female	National
Managers	1.2	0.5	1
Professionals	4.5	5.4	4.8
Technicians & associate professionals	2	1.1	1.7
Clericals support workers	1.8	1.1	1.6
Service and sell workers	19.7	8.7	16.5
Skilled agricultural, forestry & fisheries	33.6	44.5	38.6
Craft & related trade workers	15.8	23.7	18.1
Plant & machine operators	7.1	2.7	5.8
Elementary occupations	14.0	12.3	13.5
Other occupations	0.3	0.0	0.2

Source: Bangladesh Bureau of Statistics, Labor Force Surveys 2013

Table 4 shows that most of the people in the country are own account workers where 40.6 percent are following employee. Own account worker is a mainly male dominating sector than females. Women are mostly involved as family helper in Bangladesh.

Rural people are more involve in own account worker than urban and urban people are more involved as employee than rural people. Interesting thing is that rural people involved more as family helper 21.3 percent than urban 9.9 percent.

Table 4 Status of employment by area & sex

Status in employment	Rural	Urban	Male	Female	National
Employer	0.6	1.6	1.2	0.1	0.9
Own account worker	42.7	35.3	52.2	12.3	40.6
Family helper	21.3	9.9	5.1	50.1	18.2
Employee	34.3	50.3	41.2	32.8	38.7
Others	1.1	2.9	0.3	4.7	1.6

Source: Bangladesh Bureau of Statistics, Labor Force Surveys 2013

Figure 4 shows the distribution of self-employed persons by years. It shows that the number of self-employed person is increasing since 2002-03 and it is continuing to increase. In 2002-03 it was 19.8 percent and now in 2013 it is 23.6 percent.

Figure 4 Distribution of self-employed person by years

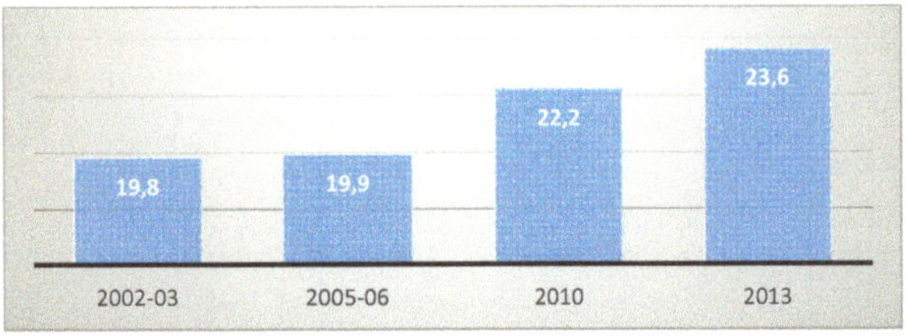

Source: Bangladesh Bureau of Statistics, Labor Force Surveys 2013

Trend of youth labor force in million below the figure 5 shows that both male and female is increasing overtime. There is a significant change which is growing rate of females are higher than the male.

Figure 5 Trend of youth labor force in Bangladesh

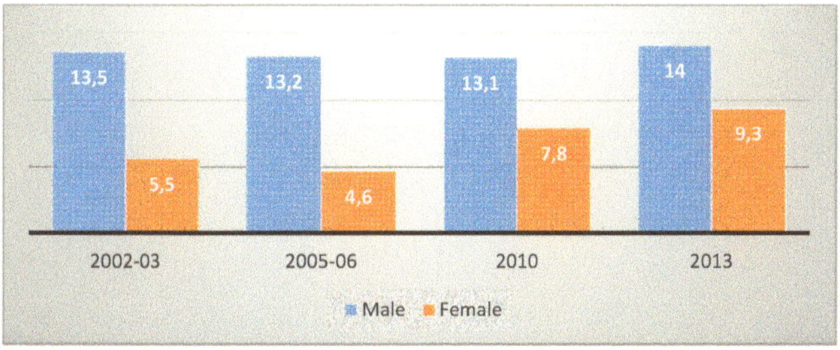

Source: Bangladesh Bureau of Statistics, Labor Force Surveys 2013

2.2. Unemployment Situation of Bangladesh

Bangladesh Bureau of Statistics (BBS) conducted Labor force survey 2013 where it adopted the definition of 'Unemployment' is given by International Labor Organization (ILO) that a person is considered as unemployed if he/she do not work at all during a week (even an hour) and is actively looking for work or is available for work but does not work due to temporary illness or less availability of job including his age should be 15 years and over.

The data shows that 4.3 percent of the economically active population is unemployed. Women are more unemployed than male in Bangladesh. The trend shows that unemployment rate for female is increasing overtime where it is decreasing for male over the years. According to BBS 2013 the rural people were more unemployed 5.7% than urban 3.7%.

Table 5 Trend of unemployment in Bangladesh

Category	**2002-03**	**2005-06**	**2010**	**2013**
Bangladesh	4.3	4.3	4.5	4.3
Male	4.2	3.4	4.1	3.0
Female	4.9	7.0	5.8	7.3

Source: Bangladesh Bureau of Statistics, Labor Force Surveys 2013

Figure below shows that unemployment rate is higher in rural areas than the urban area which is also higher than the national figure.

Figure 6: Unemployment rate by area

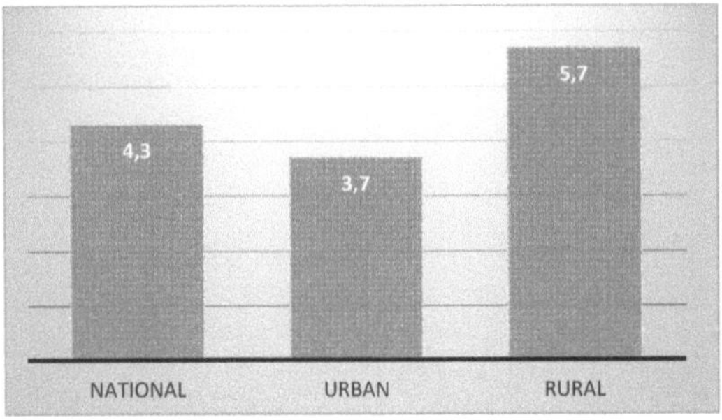

Source: Bangladesh Bureau of Statistics, Labor Force Surveys 2013

Figure 7 shows that unemployment was highest among higher secondary level 7.9 percent followed by tertiary level 6.7 percent. For less educated groups (primary, none and others) unemployment were less than then the national average.

Figure 7 Unemployment rate by education level

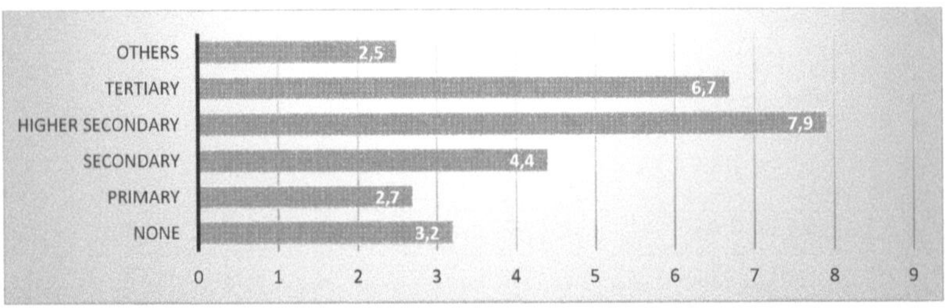

Source: Bangladesh Bureau of Statistics, Labor Force Surveys 2013

According to Labor force survey Bangladesh 2013 underemployment occurs when a person works less than 40 hours in the reference week. In 2013 2.3 million employed person 4 percent were underemployed. Among them 1.6 million were male and 0.7 million were female. Majority of them lived in rural areas around 1.8 million compare to 0.5 million in urban areas. The survey shows that the underemployment was higher among young than other age category.

The labor force participation rate of Bangladesh showed that men were more economically active employed & their labor force participation rate is higher than women which should be addressed to achieve demographic dividend in the context of the country. Although female labor force in Bangladesh has significantly increased over the last two decades. But most of them are involved in paid labor force. This rapid increase in the participation of women in labor force can be termed as feminization of labor which is a product of the increasing number of textile and garment industries (Titumir, 2013). Again most of the people are involved in low paid jobs because of their emergency (Rahman, 2013). The supply of labor is higher than the demand which leads low paid jobs in Bangladesh. Most of the people in Bangladesh is involves in agricultural sector but industrial & service sectors are two potential sectors among the economic active population. The growing trend of non-agricultural sector involved labor force which is higher in number than the agricultural sector involved people is an indicator that the country is now shifting its economy to agricultural based economy to industrial & service based economy day by day.

The labor force of Bangladesh is mainly based on informal sector. According to Ali (2013) the informalization of labor market are associated with multiple reasons. Most of them are being exploited from formal sector, job crisis and due to their lack of skill. His research also found out that lack of education and social networking are also responsible for this. Political connectivity of government or opposition party is also a vital factor for being involved in informal sector. Largest share of the employed population of Bangladesh is involved in skilled agricultural, forestry & fisheries sector where women are more involved than male. Females are mostly family works in the context of Bangladesh. Due to economic, social, cultural and political perspective of gender discrimination and lack of inequality are the main reasons for this (Islam, 2014). Lack of training & education were found two crucial factors for women not involving in formal & paid economy. According to World Bank (2011) .The growing trend of own account worker that means self-employed with no employee has reduced poverty in Bangladesh. The increasing rate of unemployment in rural area is an issue to be addressed.

According to the Quarterly Labour Force Survey Bangladesh 2015-16 done by the Bangladesh Bureau of Statistics, rural areas have 18.2 lakh unemployed people, more than double the number urban areas 7.7 lakh. Lack of access to quality education, training and technical support may be the reasons behind it. In Bangladesh, unemployment increases with the increased level of education. Higher secondary passed population are mostly unemployed following tertiary passed people. The lack of quality, applied & skilled based approach of education may be the issues in the context of Bangladesh for both unemployment & underemployment.

3. Potential Sectors of Employment in Bangladesh

3.1. Small and Medium Enterprises (Smes)

Small and medium enterprises (SMEs) both are considered as the engines of growth of innovation around the world. They play significant role for driving economic growth and creating jobs. According to Rahman (2015) Small and medium enterprises are playing a significant role for accelerating industrialization, economic growth, job generation and reducing poverty in Bangladesh. The total contribution of SMEs to export earnings differs from 75 percent to 80 percent. The industrial sector contributes 31 percent of the total gross domestic product (GDP) most of which is coming from these SMEs. According to him the total number of SMEs in Bangladesh is estimated to be 79,754 among them 93.6 percent are small and 6.4 percent are medium. The 2003 Private Sector Survey estimated that there were around 6 million micros, small and medium enterprises running with less than 100 employees. Most of the SMEs are located outside the metropolitan areas of Dhaka and Chittagong. Surprisingly SME sector of Bangladesh has created 15 lakh new jobs between 2009 and June 2014. A developing country which is experiencing demographic dividend like Bangladesh, SMEs can be considered as an effective way for the reduction of unemployment. Seasonal unemployment underemployment can be solved by the extensions of SMEs. SME can reduce the rural to urban migration in the capital Dhaka and other cities by increasing living standard of the rural population. SMEs can reduce the urban – rural income gap in Bangladesh which may be also considered as a poverty reduction strategy. Fatima and Ahmed (2017) stated that access to finance and training are two vital constraints for SMEs. The also claimed the government's policy and financial- technical support are needed to promote the SME product for the global market. The demand for SME products is still high outside Bangladesh which is a big opportunity for the country.

3.2. ICT Sector

Focusing on ICT as one of the highest priorities for the government of Bangladesh is nothing new. Since 1997 it was officially recognized that this sector can make a significant impact for the economic development of the country. In 2015 the government introduced a guideline for ICT sector. ICT sector of Bangladesh mainly do web design, mobile application development, customized web application development, BPO, web development, graphic design, application maintenance, customized software development, cloud computing system integration. It is estimated that within five years ICT sector will contribute 1 per cent to the Bangladesh's total GDP share and will produce 150,000 ICT professionals (Sadia, 2016). Now days ICT industry are seen as one of the top sectors for graduate employment in terms of creating high quality jobs and also reducing underemployment. According to Nyenrode Business Universiteit (2012) the monthly income an ICT employee is approximately US$ 200 per month in average. The study of Nyenrode Business Universiteit showed that the significance of the ICT sector for the Bangladeshi economy is increasing day by day. According to their estimations within the next five years 1 percent of the country's total GDP will come from the software and IT service based sector and over 150,000 Software and IT professionals will work in IT sector in Bangladesh.

3.3. Vocational Training and Overseas Employment

Vocational training enhances career development and employment. In Bangladesh the VET is offered at secondary level of education and especially to the students of underprivileged background (Newaz, 2013) According to Alam (2010) usually students enter the vocational training from 9th to 12th grade. The duration of the courses ranges from three to six years, depending on the type of training provided. For the largest number of non-formal trade courses, the VET begins after eight years of schooling. According to Newaz (2013) "There are overall 121 government institutes and 1,473 private institutes for VET. The total numbers of seats available in these institutes are 120,500.In 2005-2006, number of registered students in S.S.C Vocational courses was 1, 00,865. Out of these students, the number of students who attended exams were 47, 521 and the number of pass outs was 29, 299. So it shows that 61.65% of the students attending the exam passed". The government of Bangladesh has introduced two separate policies which have relevance to skills development and training Non-Formal Education Policy 2006 and National Skills Development Policy 2011. These policies are introduced for the objectives and strategies regarding different aspects of formal and non-formal education and training opportunities for economically active population (MOE, 2011).

International migration creates employment especially for trained migrants and it is the second main source after Garment sector for Bangladesh. The total number of exported workers of Bangladesh was 8.7 million and the volume of remittance inflow to Bangladesh stood at 116566 million US dollar since 1976 to 2013 (Syed, 2014). Bangladesh has become as one of the largest remittance receiving countries of the world actually 8th largest which is a great surprise (World Bank,2012).Because of shortage of labor supply in Middle East and developed country the demand of labor supply from developing country is increasing day by day. In 1976 only 6087 workers were exported from Bangladesh in 10 to 15 countries. At present its labor is exported to 159 countries of the world. (Bangladesh Economic Review, 2013).Major destination countries of Bangladeshi migrant's workers are Saudi Arab, UAE, Malaysia, U.K, U.S.A Oman, Kuwait, Singapore, Qatar and Bahrain where professional and skilled migrant worker's destination countries are mainly the U.K, U.S.A, Australia, Germany, Japan and Italy. Vocational & skilled based applied training can increase the demand of exporting man power from Bangladesh & the skilled man power can bring more remittance than the unskilled low paid workers.

3.4. Employment in RMG Sector

Bangladesh can produce huge number of jobs in RMG sector for through improving the productivity and social and environmental compliance, according to a new study by the World Bank (2016). This is supported by another study of World Bank in 2012 that 90 percent of the workers in garments were female. In 2010 about 35 percent of the employed woman worked in non-agriculture sector among them more than a one third is engaged in RMG sector (World Bank 2012).The study of World Bank (2016) said that "China is the largest apparel supplier having 41 percent global market share, Bangladesh is the second largest apparel supplier with 6.4 percent share. Bangladesh is likely to benefit from the shifting of work orders from China, as China is losing its market share due to higher cost of production". In 2000-01 number of garment industry was 3480 which increased to 5150 in 2010-11. In the same period employment in this sector was 1.5 million which increased to 3.6 million (Reza & Salim, 2011).Bangladesh should use this sector for creating quality jobs & products to increase its GDP.

3.5. Policy Suggestions

To reduce unemployment, underemployment and generate employment opportunities and for solving problems related to services the following policies are suggested.

1. Educated unemployment is a huge problem for Bangladesh. Quality education with applied specific approach rather than only theory based knowledge is necessary.

2. The quality of the higher educated in higher secondary and tertiary level is necessary. The government should take proper steps to maintain standard of degrees.

3. A uniform equal opportunity for all should be ensured by the government to reduce discrepancy in recruitment policy & rules among the different organizations regarding gender, educational qualification & recruitment process.

4. Age of entry into jobs is also discriminatory in Bangladesh. In Bangladesh the job seeking age is fixed at 30 years. Many meritious young job seekers are deprived of due to this rule of service age. The age of entry level should increase.

5. The government should consider about the expanding higher education with unemployment that's why vocational and ICT based training should be expanded instead of higher education.

6. Quota system should be properly used where it is necessary. Because of this system the meritorious job seekers are deprived of employment. But women, disable population& other vulnerable people should give priority.

7. Disable population friendly communication and transport should be created to ensure their participation into labor force.

8. Working environment and labor rights with policy from government should be ensured for justice among the population.

9. Domestic and foreign investment has to be increased to expand economic sector.

10. Political violence should be eliminated to increase investment & favorable employment situation.

11. Labor intensive production technology should be adopted to increase production job sector.

12. Technical job oriented education system should be introduced.

13. To increase the productivity & highest potential of the economic active population their health hazard and nutrition should be given priority.

14. Micro credit program should be increased for self-employed persons.

15. Recruitment process for job of Bangladesh government is so lengthy and time consuming. It should be terminated within short time period.

16. For increasing overseas employment skilled and trained work force is the key to catch the global market demand.

17. Urban-Rural education & employment discrepancy should be eliminated to reduce unemployment and underemployment in Bangladesh.

4. Conclusion

Labor force survey 2013 showed that around 41% of youth were not in employment, education or training and the portion of young unemployed was 78% in Bangladesh. The country's population under 30 years of age contributes 60% of the total labor force. A developing country like Bangladesh which is experiencing demographic dividend should grab the opportunity to take its full advantage by bringing its economically active population into labor force. The lack of quality education and skilled labor force are two main reason of unemployment including underemployment in the context of Bangladesh without any doubt. Lack of investment into economy especially foreign investment is another cause for unemployment in the country. Lack of exporting man power is also a vital factor. Eradicating disparities among gender and region should be addressed by policy perspective. Bangladesh has to take its first demographic dividend's full advantage transforming them into human resource ensuring their employment including rights, dignity & prosperity.

5. References

Ali, A.M. (2013). Informal Labour Force, *Unnayan Onneshan*, Dhaka, Bangladesh.

Alam, Z. (2012). Bangladesh: Development, Dreams and Demographic dividends, Retrieved from https://mygoldenbengal.wordpress.com/2012/11/19/bangladesh-development-dreams-and-demographic-dividends/

Alam, G. M. (2010). A proper plan needed in place for upcoming technical, vocational, education projects. *Journal of University of Malaya*, 2(3), 72-83

Bangladesh Economic review, (2013). Ministry of Finance, Government of the people's republic of Bangladesh. Retrieved from https://www.mof.gov.bd/en/index.php?option=com_content&view=article&id=301&Itemid=1

Bangladesh Bureau of Statistics (BBS) (2015). Labour Force Survey Bangladesh 2013, Statistics and Informatics Division, Ministry of Planning, Government of the people's republic of Bangladesh

Bangladesh Bureau of Statistics (BBS) (2017). Quarterly Labour Force Survey Bangladesh 2015-16, Ministry of Planning, Government of the people's republic of Bangladesh

Bangladesh Bureau of Statistics (BBS) (2011). Population and Housing Census 2011. Statistics and Informatics Division (SID) Ministry of Planning. Government of the people's republic of Bangladesh

Daniels, L. &International Consulting Group (2003). National Private-Sector Survey of Enterprises in Bangladesh, 2003. Retrieved from https://www.jobstrust.org/jobsproject/programs/Sl%2033.%20Natinal%20Private%20Sector%20Survey.pdf

Fatima, K. & Ahmed, S. (2017 March 19).An alternative 'driving force' of economic growth, The Independent, Retrieved from http://www.theindependentbd.com/printversion/details/86089

Islam, M.S. (2014). Women's Empowerment in Bangladesh: A Case Study of Two NGOs, Bangladesh Development Research Working Paper Series (BDRWPS) No 23

Ministry of Education. (2011). National Skill Development Policy. Dhaka. Ministry of education

Nyenrode Business Universiteit (2014). ICT Sector Study Bangladesh Bridging the Gap between Dutch and Bangladeshi ICT sectors. Netherlands Bangladesh Business Platform, January 2014

Newaz, M.T. (2013). Vocational education and training in Bangladesh: Why it is not working? *International Journal of Research Studies in Education* 2013 October, Volume 2 Number 4, 29-40

Nabi, A.K.M.N. (2011). Demographic Trends in Bangladesh, Retrieved from http://www.iedcr.org/pdf/files/NPHC%20WEB/UPLOAD-4/Dr.Nurun%20Nabi.pdf

Population Reference Bureau (2016). World Population Data Sheet, Retrieved from www.prb.org/pdf16/prb-wpds2016-web-2016.pdf

Rahman, K.M.M. (2013). Women in Labour Force, *Unnayan Onneshan*, Dhaka, Bangladesh

Rahman, M.F. (2015 March 11).Potential of Small and Medium Enterprises, The Daily Star , Retrieved from http://www.thedailystar.net/supplements/24th-anniversary-the-daily-star-part-2/potential-small-and-medium-enterprises-70927

Rahman, M. (2017 March 19).The SME sector has the highest opportunity to create employment in Bangladesh, The Independent, Retrieved from http://cpd.org.bd/sme-sector-highest-opportunity-create-employment-bangladesh-mustafizur-rahman/

Reza, M., Selim (2011) Impact of global financial crisis on the Economy of Bangladesh, *Bangladesh Journal of Political Economy- BEA.*

Sadia, F. (2017). Exploring prospects of ICT in Bangladesh Retrieved from http://forum.daffodilvarsity.edu.bd/index.php?topic=42413.

Syed, A.S. (2014).Export of Manpower: Problems and Prospects, *Monthly Aikar Barta*, September and October 2014.

Titumir, R.A.M. (2013). *Compulsive Structures and Embedded Institutions*, in Titumir, R.A.M. (ed.), Institutions Matter: State of Women in Bangladesh, 2013. pp. 1-14. Unnayan Onneshan, Dhaka, Bangladesh.

UNFPA (2014). The State of World Population Report 2014: The power of 1.8 billion: adolescents, youth and the transformation of the future. Retrieved from https://www.unfpa.org/sites/default/files/pub-pdf/EN-SWOP14-Report_FINAL-web.pdf

World Bank (2011). More and Better Jobs in South Asia, 1818 H Street NW Washington DC 20433

World Bank (2012). Bangladesh Development Report, Vol. 2. Retrieved from https://siteresources.worldbank.org/INTWDR2012/Resources/7778105129969968583/7786210-1315936222006/Complete-Report.pdf

World Bank (2016). Bangladesh Development Update, Sustained Development Progress, The World Bank Office, Dhaka.

World Bank (2016). Stitches to Riches? Apparel Employment, Trade, and Economic Development in South Asia, Gladys Lopez-Acevedo and Raymond Robertson, Editors, 1818 H Street NW, Washington, DC 20433